OIL

Graham Rickard

Thomson Learning
New York

Titles in this series

Bricks
Electricity
Gas
Glass
Oil

Paper
Plastics
Steel
Water
Wood

Cover: (Main picture) Oil workers in the Middle East getting ready to drill. (Top right) Light oil pours onto a machine part to make it run smoothly.

First published in the
United States in 1993 by
Thomson Learning
115 Fifth Avenue
New York, NY 10003

First published in 1992 by
Wayland (Publishers) Ltd

Copyright © 1992 Wayland (Publishers) Ltd

U.S. revision copyright © 1993 Thomson Learning

Cataloging-in-Publication Data applied for

ISBN: 1-56847-045-2

Printed in Italy

Contents

What is oil?	4
How oil was formed	6
The history of oil	8
Sources of oil	10
Finding oil	12
Drilling for oil	14
Oil under the sea	16
Refining crude oil	18
Transporting oil	20
Using oil	22
Chemicals from oil	24
Oil pollution	26
Projects with oil	28
Glossary	30
Books to read	31
Index	32

All the words that appear in **bold** are explained in the glossary on page 30.

What is oil?

In its natural form, oil is a thick black or dark brown liquid, which is found underground in many parts of the world. This natural substance is called crude oil and contains two main **elements**, hydrogen and carbon. Oil has been used for at least 2,000 years, but it is only since the nineteenth century that it has become the most important of all the world's **minerals**.

In its natural form, crude oil is a thick, sticky liquid. It is the most valuable of the world's fossil fuels.

Without oil, our modern way of life would be impossible. Oil provides **fuel** for transportation, and it heats our homes and offices. It generates electricity, and it is essential for heat processes in industry. Oil products are also used as **lubricants**, substances that keep our machines working smoothly. Chemicals from oil are used to make a wide range of products, including plastics. Oil is so valuable to us that it is sometimes called "black gold."

Mobile cranes are used to build an oil pipeline across the desert in Yemen, Arabia.

How oil was formed

Oil is a **fossil fuel** and was formed millions of years ago from the remains of tiny sea creatures. When these creatures died, their bodies fell to the sea floor and were then covered by the mud and sand that had been washed into the sea by rivers.

Over millions of years, the mud and sand became harder and their weight turned the remains of the sea creatures into tiny drops of crude oil. The oil that formed in loose-grained rocks, such as

Millions of years ago, dead sea creatures sank to the sea floor. Mud and sand buried them and hardened them into rock.

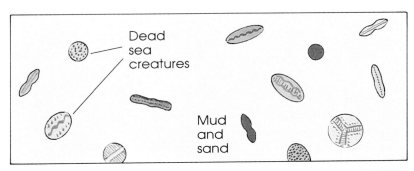

Millions of years later, the weight of the rock has turned the bodies of sea creatures into drops of oil.

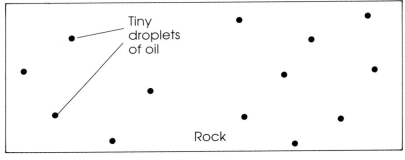

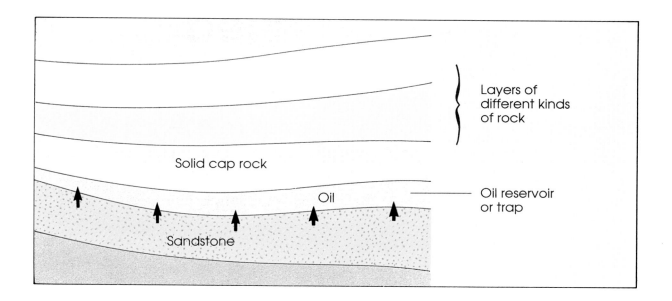

sandstone, was squeezed upward by pressure from the weight of the rocks above. The oil was trapped by a different type of rock, called a cap rock, which it could not pass through.

Where there was no cap rock, the oil sometimes came to the surface and formed **seepage**. But where the cap rock was large, enormous amounts of oil could become trapped in the **permeable** rocks below the cap rock—just like water is trapped in a sponge. These oil traps are called **reservoirs**, and it is from these deep underground traps that we now obtain our oil.

The drops of oil that filled the holes in spongelike rocks, such as sandstone, were squeezed upward and trapped beneath more solid cap rocks to form an oil reservoir.

The history of oil

In some parts of the world, crude oil seeps through the earth's surface. People have been making use of this oil supply for thousands of years.

In ancient China, oil was used in simple lamps, while Native Americans used it as a medicine. Crude oil was sometimes found as a black, sticky form called pitch, which was used as a waterproof coating more than 2,000 years ago.

All through history, people have made use of the oil that seeps up to the earth's surface.

For many centuries, crude oil has been obtained from wells dug by hand. But the modern oil industry really began on August 27, 1859, when Colonel Edwin Drake drilled an oil well in Pennsylvania and discovered oil at a depth of 69 feet. This started an "oil rush" in the area. People soon started discovering oil and drilling for it all over the world. Inventions like the automobile and its **internal combustion engine** led to a huge demand for oil. Today, the oil industry produces 60 million **barrels** of oil every day.

The world's first drilled oil wells were in Pennsylvania. This oil was stored in wooden barrels and was transported on riverboats.

Sources of oil

Oil is found on every continent except Antarctica, but some parts of the world produce more oil than others. This is mainly because of the types of rock that lie deep underground in each region. After the first discoveries of oil in the United States, oil was later found in Southeast Asia, Russia, and the Middle East. Many companies were set up to produce, transport, and sell this oil.

The world's main oil-producing areas are the Middle East, Russia, and the United States.

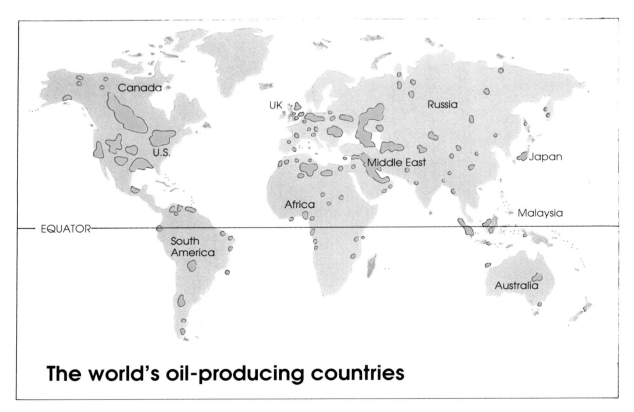

The world's oil-producing countries

Two workers fit the drill shaft into an oil rig in Mexico.

The Middle East is now the world's largest source of oil, but oil is also produced in the United States, Europe, Mexico, Russia, and Venezuela in South America. New supplies of oil are still being found all over the world. China now produces large amounts of oil, and Egypt and Thailand have recently become oil producers.

Oil companies are always looking for new sources of oil. New reserves are being found in such unlikely places as the North Sea and Alaska. But today we are using oil faster than we are finding new sources.

Finding oil

In a seismic survey in Nigeria, engineers place a stick of dynamite into the ground.

Finding oil is a mixture of science and guesswork. In the early days of the industry, finding oil was a matter of luck, but since then, finding oil has become much more scientific.

Geologists are scientists who study the structure of the earth. They use several different methods to find oil. They study photographs taken from airplanes and **satellites**, looking for the right kind of rock structures. They also explore remote areas and take samples of rocks to see if they contain any oil or hold any clues as to where oil might be found.

Explosions are used in seismic surveys to send shock waves through the ground. Special instruments are used to measure the echo of these shock waves.

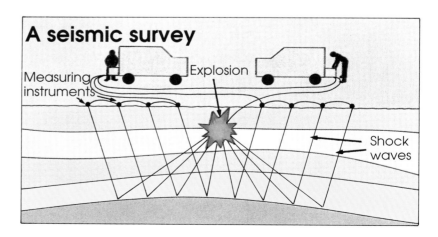

Seismic surveys are carried out by geologists to find oil reservoirs. They involve making a small explosion just below the ground and taking measurements of the **shock waves** that travel through the underground rocks.

In spite of all these modern scientific methods, finding oil is still a difficult and expensive job, and the only way to be sure of finding oil is to actually dig a hole down into the ground.

A small, mobile drilling rig is used to dig bore holes in a seismic survey in Utah.

Drilling for oil

As the drill digs deeper into the ground, new sections of pipe are attached to the drill shaft.

When geologists decide that there might be oil in a certain area, the next step is to drill a test well.

To drill for oil, a hollow revolving **drill bit** is pushed down through the rock. These drill bits have teeth made of diamond or especially hard materials that can cut through rock like sandstone at the rate of 100 yards per hour. The drill bit is attached to the bottom of a steel pipe, and as each piece of pipe sinks into the ground, more lengths are added as the bit travels farther down.

"Rocking horse," or "nodding donkey," oil pumps are a common sight in the United States. This one is decorated to look like a grasshopper.

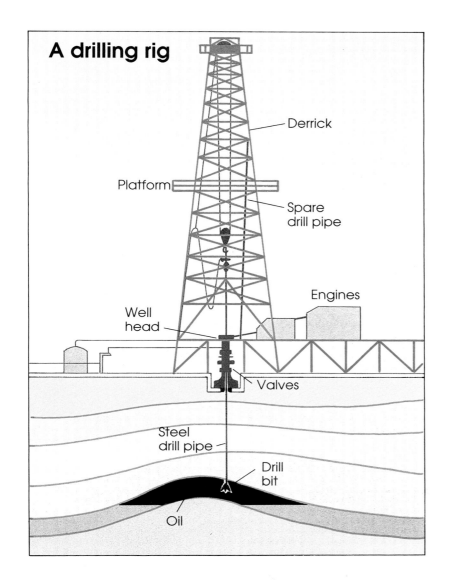

An oil rig sits on top of an oil well and supports the long drill bit as it cuts into the rock below.

If the bit cuts into a rock that contains oil, oil and gas rush up the pipe at high pressure. They are contained by special **valves** at the well head. The gas is removed, and the oil is stored in tanks or is pumped through a pipeline to storage areas or to a **refinery**.

Oil under the sea

An oil rig in the stormy waters of the North Sea.

There are three main types of offshore oil rigs.

Finding oil beneath the sea is similar to finding oil on land, but it is much more difficult, expensive, and dangerous.

The first underwater oil well was drilled in 1897 off the coast of Southern California, but it is only since the 1970s that oil wells have been sunk offshore at sea.

The first offshore drilling was done by using **rigs** fixed to the sea floor, but these rigs could only drill in one place. For test drilling, a movable rig with legs that could be lowered onto the sea floor was invented. When drilling is finished, the legs are raised and the rig is towed to a new site.

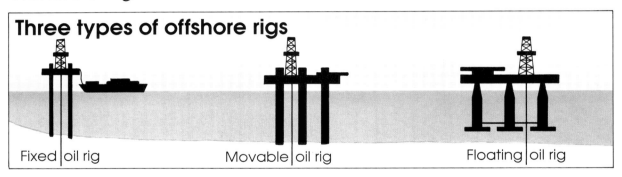

Three types of offshore rigs

Fixed oil rig — Movable oil rig — Floating oil rig

When a well starts to produce oil, a production platform is built to house all the machinery and oil workers. The oil from the well is sent ashore by ship or by underwater pipeline. Food and equipment for the crew are usually flown to the platform by helicopter.

Helicopters are often used to transport oil workers and materials to and from the oil rigs.

17

Refining crude oil

Crude oil is of very little use to anyone, and it has to be treated in several different ways at an oil refinery before it can be used. Crude oil is a mixture of many different gases, liquids, and solids and contains small amounts of impurities, or unwanted chemicals. All these substances have to be separated during the refining process.

The crude oil is heated in a tall tower divided into separate parts.

Many oil refineries are built on the coast, and crude oil is delivered to them by large tankers.

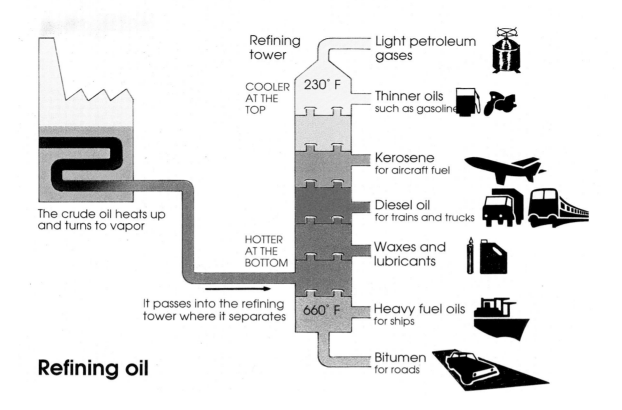

Refining oil

The lighter fuels, such as gasoline, evaporate first and are collected from the top when they cool. The heavier parts of crude oil such as diesel fuels are collected at the bottom of the tower.

Refining crude oil results in many useful products. These include butane gas, gasoline, diesel fuel, kerosene fuel for aircraft, and bitumen for making roads. As well as fuels and lubricants, refining produces many other valuable chemicals, called **petrochemicals**, which can be made into plastics.

This diagram shows how crude oil is split into gases and oils at different temperatures in the refining tower.

Transporting oil

Tanker trucks can carry thousands of gallons of fuel.

Every year, huge quantities of oil and oil products have to be moved from oil wells to refineries, and from refineries to garages, factories, and power stations.

When oil was first produced at the end of the nineteenth century, it was stored in barrels and was transported on horse-drawn wagons, steam trains, and cargo ships. Today, large amounts of oil are often pumped along very long pipelines. The world's longest pipeline is the Trans-Siberian in Russia. It runs for 2,318 miles.

This huge pipeline carries oil across the frozen oil fields of Alaska to tankers that transport it to refineries.

Modern oil tankers are so enormous that the crew sometimes use bicycles to travel from one end of the deck to the other.

In North America, another pipeline runs for 1,771 miles, from Alberta, Canada, to Buffalo, New York, in the United States.

Oil products are needed all over the world, and large amounts of oil have to be transported across the oceans in oil tankers. Some of the tankers are huge—the *Seawise Giant* weighs over 564,500 tons and is about 1,500 feet long.

Using oil

Without oil, our modern world could not exist. Our industries would stop working, and all our machines and transportation would grind to a halt.

Most of the world's oil is used as fuel to power our cars, trains, and aircraft, and to provide us with heat and light.

This diagram shows how much oil was used around the world in one year.

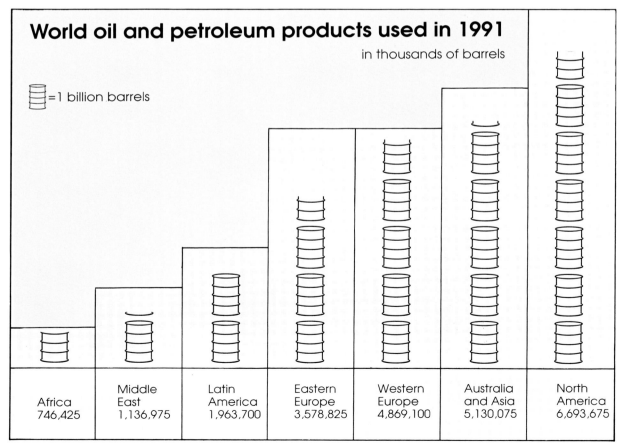

World oil and petroleum products used in 1991

in thousands of barrels

= 1 billion barrels

Africa	Middle East	Latin America	Eastern Europe	Western Europe	Australia and Asia	North America
746,425	1,136,975	1,963,700	3,578,825	4,869,100	5,130,075	6,693,675

Without oil to lubricate it, this machine would not be able to cut the metal to make propeller blades for ships.

In many countries, electricity is produced by burning oil in power plants, and many industries burn oil to provide the heat to make their products. In some parts of the world, kerosene, or paraffin, is still widely used for lighting, heating, and cooking.

All machines need lubricants to make them work smoothly and last longer. Lubricants work by reducing the **friction** between two surfaces that are rubbing against each other. Thousands of different types of lubricants can be made from crude oil, from very thin oil for watches to thick grease for heavy machinery.

Chemicals from oil

Today, plastics are used to make a huge variety of things.

Paint is made from plastic solvents, which come from oil.

Some light oils are separated from crude oil and are sent to a chemical factory. About 10 percent of the world's crude oil is now made into petrochemicals, which are made into a wide range of materials.

Many types of plastics are made from oil. Things we use everyday—buckets, toothbrushes, appliances, bowls, and toys—are made from plastics. Plastics are also made into paint solvents, glues, and packaging. Other types of plastics give us synthetic materials such as nylon, rayon, and acrylics.

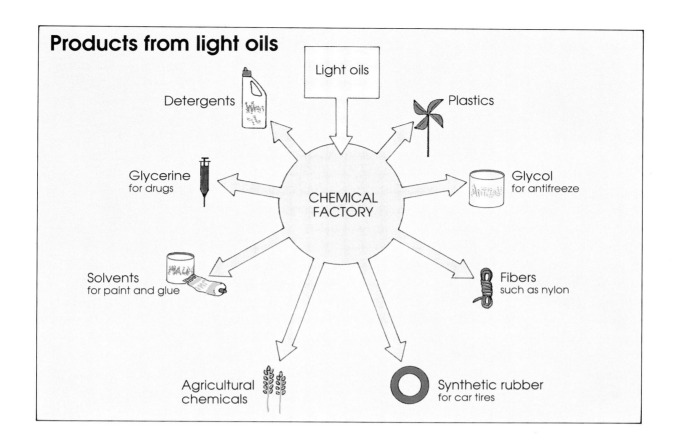

Synthetic rubber can also be produced from plastic. It is used to make car tires and hard-wearing soles for shoes.

Many other oil-based chemicals are used for different purposes. Petrochemicals can be made into **detergents** for washing dishes; agricultural chemicals for fertilizing the soil and destroying pests; **glycerine** for drugs; feed for animals; **cosmetics**; and explosives.

Chemicals that come from crude oil can be made into a wide range of everyday products.

Oil pollution

Although oil is a natural product, our use of it can cause damage to the **environment**.

Like coal and gas, when oil is burned it gives off gases into the air. In power plants, these gases are sent high up into the **atmosphere** and can be blown hundreds of miles before falling in the rain. This rain is known as "acid rain" and can kill trees and harm fish in lakes and rivers.

Gas from oil wells is usually burned. This may be harmful to the environment because it increases the amount of carbon dioxide in the atmosphere.

A gas called **carbon dioxide** is also produced when oil, coal, or gas is burned. This increases the amount of carbon dioxide in our atmosphere. Many scientists are worried that this will add to the greenhouse effect and will make the earth warmer, resulting in damage to the environment.

From time to time, oil is accidentally spilled and causes pollution at sea. Some of these oil slicks are very large, such as the slick caused by the tanker *Exxon Valdez* that ran aground off the coast of Alaska. A spill of oil kills fish, birds, and other sea life.

Above *in Mexico City, fumes from cars and factories create an unhealthy fog.*

Below *Seabirds suffer when an oil tanker spills its load.*

Projects with oil

How oil fields are formed

You will need:

Coarse cleaning sponge
Water

A clear saucer
A large bowl

In this experiment, the sponge represents the permeable rock, and the water is the oil in the pores of the rock. The saucer represents the impermeable cap rock.

1. Fill the bowl halfway with water. Float the sponge on top of the water.

2. Press the sponge into the water by using the saucer. Where has the water collected?

You should find that the water drawn up in the sponge collects underneath the saucer. The water cannot pass through the saucer—just as it cannot pass through the impermeable cap rock in the earth—so it collects beneath it and forms a "reservoir."

Oil and detergents

You will need:

A large bowl
Water

Cooking oil
Dish detergent

When oil spills occur, they are often cleaned up with detergents. This experiment shows why this is possible.

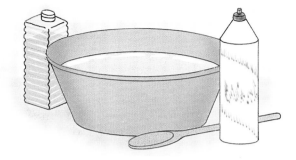

1. Partly fill the bowl with water, and pour in a teaspoonful of cooking oil. Try to mix the oil and water, using a spoon. Will they mix together?

2. Squirt some dish detergent into the bowl, and stir the mixture thoroughly. What happens to the oil?

You should find that the detergent breaks up the oil, and the oil now mixes with the water and disappears.

Oil as a lubricant

You will need:

A baking tray
A large coin

Cooking oil
A protractor

1. Place the baking tray flat on the table, with the coin on top of one end of the tray.

2. Stand the protractor upright beside the tray at the opposite end to the coin.

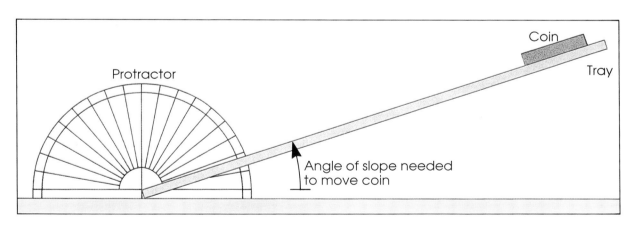

3. Slowly lift the tray at the end that has the coin on it. Note the angle of the tray by measuring it on the protractor. Write down the angle when the coin begins to slide.

4. Coat the baking tray with cooking oil and repeat the same process. The coin will start to slide sooner. What is the angle this time? Why do you think this is?

Glossary

Atmosphere The layer of gases that surrounds the earth.
Barrel A type of container. Each barrel in the oil industry contains 40 gallons of oil.
Carbon dioxide A colorless gas that is in the air.
Cosmetic Something that we use to improve our appearance.
Detergent A substance that breaks down oil and grease, such as dish detergent.
Drill bit A strong, sharp object that is screwed on to the end of steel piping and is used to bore through rock.
Element A simple chemical.
Environment Our surroundings.
Fossil fuel A fuel such as oil, gas, or coal that was formed millions of years ago from dead plants and animals.
Friction The resistance between two surfaces when they rub together.
Fuel A source of energy that is burned to provide heat or make an engine work.
Glycerine A syrupy liquid.
Internal combustion engine An engine that is powered by burning fuel, such as gasoline.
Lubricant A substance, such as oil, which reduces friction between two surfaces.
Mineral A substance that is found in the earth.
Permeable Something that lets liquids pass through it, such as a sponge.
Petrochemical A chemical that is made from oil.
Refinery A place where crude oil is broken down into different fuels and other products.
Reservoir A large store of liquid, such as oil or water.
Rig The structure that is used to drill for oil beneath the ground.
Sandstone A permeable rock consisting of grains of sand held together with minerals.
Satellite An object sent into space that sends back information to earth.
Seepage Liquid or moisture that has collected by oozing through a material such as rock.
Shock waves Vibrations that travel through the earth after an explosion.
Valve A device that controls the flow of liquids or gases. A bathroom faucet is a type of valve.

Books to Read

Asimov, Isaac. *How Did We Find Out About Oil?* New York: Walker & Co., 1980

Brice, Raphaelle. *From Oil to Plastic.* Chicago: Childrens Press, 1988

Dineen, Jacqueline. *Oil and Gas.* Hillside, NJ: Enslow Publishers, 1988

Hawkes, Nigel. *Oil.* New York: Gloucester Press, 1985

Kraft, Betsy Harvey. *Oil and Natural Gas.* New York: Franklin Watts, 1982

Lynch, Michael. *How Oil Rigs Are Made.* New York: Facts on File, 1986

Mercer, Ian. *Oils.* New York: Gloucester Press, 1988

Pampe, William R. *Petroleum: How It Is Found and Used.* Hillside, NJ: Enslow Publishers, 1984

Scott, Elaine. *Oil! Getting It, Shipping It, Selling It.* New York: Frederick Warne, 1985

Useful addresses

American Petroleum Institute
1220 L Street, NW
Washington, DC 20005

Texaco Inc
Public Relations Department
2000 Westchester Avenue
White Plains, New York 10650

Index

acid rain 26
Alaska 11, 20
Antarctica 10

cap rock 7, 28
chemicals 24, 25
crude oil 4, 6, 8, 9, 18, 19, 23

Drake, Colonel 9
drilling 14, 15, 16

electricity 5, 23
environment 26, 27
exploration 12, 13

fuel 4, 5, 22
 butane gas 19
 diesel 19
 kerosene 19, 23
 paraffin 23

gas 15

geologists 12, 13, 14
greenhouse effect 27

industry 5, 22, 23

light oils 24
lubricants 23, 29

Mexico 11
Middle East 10, 11

North Sea 11, 16

oil industry 8, 9
 pipelines 5, 15, 17, 20, 21
 products 5, 19, 20, 21, 22, 23, 24, 25
 pumps 14
 refineries 15, 18, 20
 reservoirs 7, 13
 rigs 11, 13, 15, 16, 17
 slicks 27

tankers 18, 20, 21
wells 14, 15, 16, 17, 20

petrochemicals 19, 24
pitch 8
plastics 5, 19, 24
power plants 20, 26

rocks 6, 7, 12, 13
Russia 10, 11

sandstone 7, 14
seismic surveys 12, 13
South America 11
Southeast Asia 10
synthetic materials 24, 25

United States 9, 10, 11, 14, 16

Acknowledgments

Thomson Learning would like to thank the American Petroleum Institute for checking the text for clarity and accuracy of information.

The publishers would like to thank the following for allowing their photographs to be reproduced in this book: Bruce Coleman Ltd, 8 bottom (Dieter and Mary Plage), 20 bottom (Gary Retherford), 24 top (Timothy O'Keefe); Environmental Picture Library, 27 (top); Mary Evans Picture Library, 9; J. Allan Cash Ltd, 12, 14 (bottom); Tony Stone Worldwide, *cover* (bottom), *title page*, 4, 5, 11, 17, 21, 23, 24 (bottom), 26; Topham Picture Library, 27 (bottom); Wayland Picture Library, 14 (top), 16; Zefa Picture Library, *cover* (top), 13, 18, 20 (top). All artwork is by Jenny Hughes.